HOW TO BUILD A JUNKYARD STILL

Michael H. Brown

Desert Publications
El Dorado, AR 71731-1751 U. S. A.

How to Build a Junkyard Still

©1980 by Desert Publications
215 S. Washington Ave.
El Dorado, AR 71730 U.S.A.
1-870-862-3811
info@deltapress.com

Printed in U.S.A
ISBN: 978-0-87947-301-3
12 11 10 9 8

Desert Publication is a division of
The DELTA GROUP, Ltd.

Direct all inquiries & orders to the above address.

Warning!

The Publisher (Desert Publications) produces this book for informational and entertainment purposes and under no circumstances advises, encourages or approves of use of this material in any manner.

Table of Contents

INTRODUCTION

Throughout the farm belt of the United States, many farmers plan the majority of their operations in terms of corn. "How many acres of corn should I plant?" "How many bushels will I be able to produce?" "How many hogs and cattle can I feed?"

Realistically, much of our entire agricultural economy revolves around that magic plant — corn. Corn has the capacity to reproduce itself 1,000 times from one kernel in five short months. At the same time, it will yield tons of cellulose and fiber in the plant foliage.

Not since the innovation of hybrid seed replacing open-pollinated corn has anything been viewed with so much interest by the farmer as the making of ethanol for fuel. Many farmers are able to recall the independence they had when they went to their own grain bins to get corn or oats to "fuel" their draft horses. Now, they are feeling the horrible desperation of knowing they are at the mercy of OPEC, the American oil giants, or even some as yet unheard of dictator who can obstruct the flow of petroleum into the United States.

What better solution than to produce ninety gallons of ethanol from a ton of shelled corn? From this activity, there will remain about 820 lbs. of distillers' grain which contains 200 lbs. of protein that can be fed directly to livestock. Let's look at this in terms of dollars per bushel of corn. Let one gallon of ethanol represent the energy equivalent of $1.20

worth of gasoline at present prices. 2.5 gallons of ethanol will be realized from one bushel of corn, so its worth would be $3.00. The protein is worth another 60 cents. Based on the above figures, the farmer has realized a gross return of 63% more than the present price of $2.20 per bushel of corn. He also has saved himself the interest on operating capital necessary to purchase petroleum fuels for the cropping season, because that requires cash which he normally borrows.

These are some of the reasons why we are interested in promoting the use of alcohol as fuel here at Clinton Community College. We are not enthusiastic about promoting gasahol because that is selling nine gallons of gasoline for oil interests and only one gallon of ethanol. We wish to demonstrate to the farmer the most economical and efficient ways to make this fuel, use it in his equipment and properly utilize the distillers' grain by feeding it to his livestock on his farm.

This is an ongoing project with a working committee comprised of:

    (a) The farmers on whose farms the demonstration unit will be placed.

    (b) Agri-business personnel who are welders and engine mechancics.

    (c) Five departments from the College: Agriculture, Microbiology, Physics, Drafting and Chemistry.

Clinton Community College will announce workshops to be held periodically for farmer participation as well as for other interested persons.

It is with these objectives in mind that this booklet is being printed. With the aid of Michael H. Brown as consultant, we intend to make farm ethanol production an important consideration for every farm of medium to larger size. Mike Brown has spent years studying and working with the correct principles of making alcohol, as well as the adaptation of this fuel to various motors.

Nothing found in this booklet can be disputed because everything explained here can be duplicated, tested and proved. We welcome your participation in the learning process presently and in the production process in the future.

Warren T. Bloom
Chairperson, Agricultural Department
Clinton Community College
Clinton, Iowa 52732

A standard galvanized garbage can may be used for the cooker for your junkyard still. A garbage can has a built-in safety valve in that if the pressure inside becomes too high, the lid will rise up and allow some of the vapor to escape.

Be Your Own Motor Fuel Magnate

Unlike the equipment needed to produce petroleum products such as gasoline and diesel fuel, which can run into hundreds of millions of dollars, the parts needed to construct a small alcohol still should run less than a hundred dollars, even at today's prices.

Admittedly, many of the items needed to construct this "junkyard still" are "scrounged" second-hand rather than purchased new, but the still will nevertheless enable you to brew your own motor fuel. About all you need is some old plumbing pipe, a few elbows and couplings, some radiator hose and hose clamps, a garbage can or oil drum or two, an old automobile radiator or air conditioner, some pipe nipples and the tank off an old toilet and you're in business.

In addition to the above items, you might need some gasket sealer, a monkey wrench and a screwdriver. The still isn't hard to build, but be flexible in your thinking. None of the parts listed or described herein are the ultimate in distillation engineering, so don't be afraid to substitute. After all, the state of the art of junkyard still construction is, "if it's handy and it works, use it."

The junkyard still is a crude but fairly efficient (although production is somewhat slow) device used for the thermal (heat) separation of liquids, including but not limited to the separation of alcohol from water.

In the petroleum industry this device is known as a "packed column" still. Obviously their stills are a trifle more elaborate than ours, but they both work on the same principle:

Alcohol boils at 173 degrees Fahrenheit and water boils at 212 degrees F., so a temperature control is essential to getting the alcohol out while leaving the water (about 85% of the original volume) behind. Since there is no way to control the less than 40 degrees difference in the boiling points of the two liquids with junkyard parts, we simply boil everything that moves and let the packed column do the separating after the alcohol and water have been vaporized.

Once the alcohol and water have become vaporized, they rise up through the packing material. Moonshiners commonly used rocks, but marbles or similar material can be used as the packing. The checmical terminology for these "rock stills", as they are known back in the hills, is "fractionating column".

Since the water vapor left the surface of the boiling liquid at 212 degrees F., it only needs to cool one or two degrees to recondense into a liquid and drop back down the column into the cooker. This cooling process is accomplished as the water comes into contact with the cooler surface of the packing material (rocks, marbles, etc.). The packing material absorbs heat from the water molecules, thus reducing the water vapor back into a liquid.

The alcohol vapor, on the other hand, would have to be cooled down by more than 30 degrees F. before it would recondense back into a liquid and plop back into your cooker.

This normally doesn't happen (unless you stick too much packing material into the still) and alcohol as high as 190 proof can be obtained. Of course, the alcohol must be condensed once it gets past the column, but more on that later.

Construction

So much for principle — let's bolt this junk together.

There are several ways to do this but to avoid confusion I will start with my favorite construction technique first. Doing it "this way" shouldn't take you more than an hour or two at first. With practice you will be able to bolt one together in 20 minutes.

Start with a garbage can and lid, a 3-foot length of 2 inch pipe and a pipe floor flange, some screen wire (about a square foot), a thin piece of weather-stripping to glue inside the rim of the lid, a reducing coupling for the top of the 3-foot length of 2 inch pipe, and a pipe elbow.

The garbage can serves as both fermenter and cooker. After the fermentation has taken place in your cooker for 3 or 4 days all you have to do is put the garbage can lid on, light a fire under it, and wait for your motor fuel to come out the other end.

There are a couple of modifications you have to make to the garbage can lid first. Number one, cut a 2 inch hole in the top of the lid - smack dab in the middle.

If you can obtain a garbage can that has handles on the rim of the lid also, so much the better as this will make handling the lid easier.

Next drill four holes in the lid to match the bolt holes in the pipe flange. Mount the flange to the top of the lid, sandwiching a small piece of screen between them. Use a generous amount of gasket sealer because this assembly must be air

A hole should be cut in the center of the garbage can lid and screen wire or rat wire welded in place over the hole. Drill four more holes in the lid as indicated and bolt a pipe flange to the lid as shown.

tight. This is a permanent assembly so don't worry about having to take it apart later.

The best screen to use is about one-fourth inch mesh — the kind used to rat proof corn cribs.

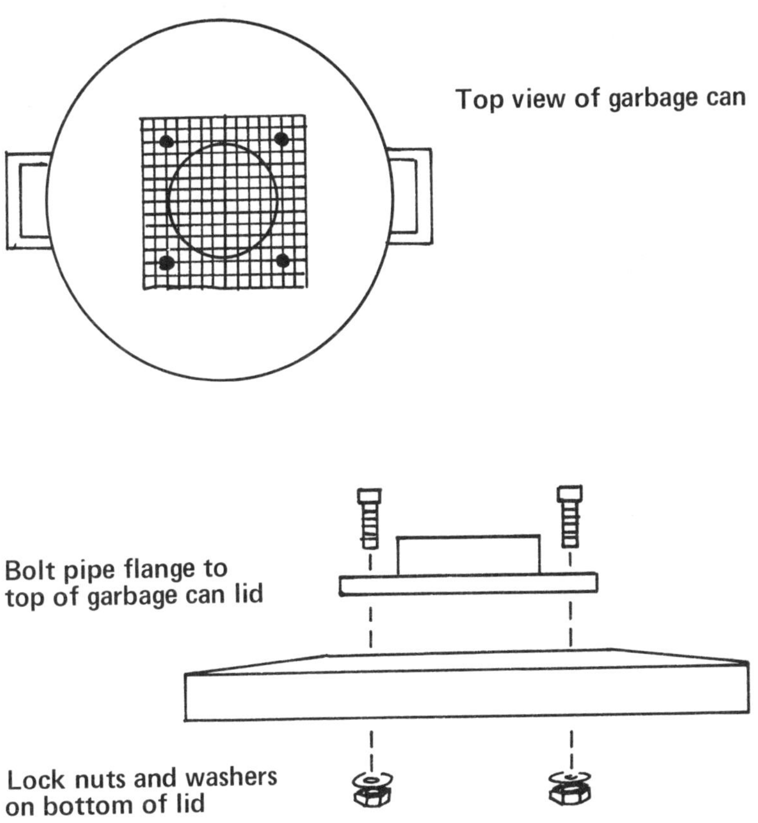

The last step with the garbage can is to glue weather stripping or something similar inside the rim of the lid so you have an air tight seal when the lid is placed on the can. Any leaks anywhere in this system and your alcohol will escape into the atmosphere instead of going where it is supposed to.

Use a generous amount of gasket sealer around bolts on bottom of lid and pipe flange on top to prevent leaks. You will not achieve a perfect vacuum with a junkyard still, but you do want to make it as efficient as possible.

The essential ingredient for your fractionating column is a 3 foot length of 2 inch pipe. Begin column assembly by screwing one end of pipe into pipe flange atop garbage can lid.

The first step in constructing your column is to simply screw the 3-foot length of 2 inch pipe into the pipe flange. Paint the threads at each end with heavey grease of pipe compound to prevent leaks.

Next, fill the 2 inch pipe with marbles or small rocks (not so small that they fall through the screen at the bottom). Back in the "old days" this was known as a "rock still."

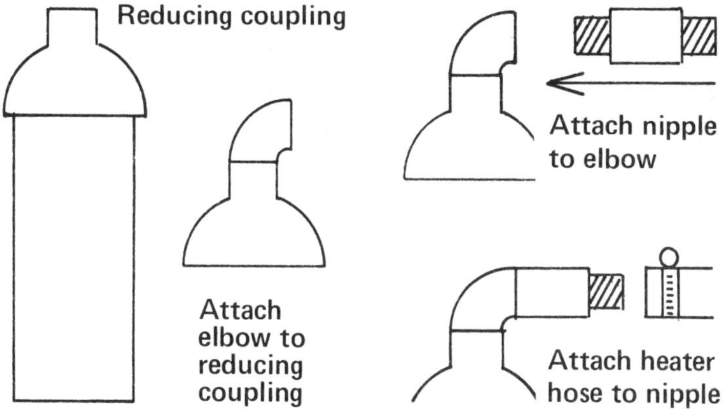

Next screw the reducing coupling to the top of the 2 inch pipe. On to this coupling goes a pipe elbow. Don't forget to grease the threads on all the pipe connections.

Use automotive heater hose (from a junk car if you want to save a few coins) to plumb up your still to the condenser.

How to build a condenser:

Because our condenser consists of a series of pipe elbows and nipples zig-zagging back and forth, an old toilet tank becomes a natural to hold the cooling water. It, also, has some holes in just the right place, and last but not least, can be scavenged for little to nothing. Naturally, it is necessary to remove all of the float valve and flushing mechanism to make room for your "plumber's nightmare" condenser.

After you have your column screwed securely into place, pour small rocks or marbles down pipe. Screen wire beneath pipe flange prevents the rocks or marbles from falling into cooker. As steam rises from the cooker into the column, the rocks or marbles help dissipate the steam's heat, causing the water vapor to recondense into a liquid and drop back down into the cooker while the alcohol vapor continues to rise.

Screw a 3/4 inch reducing coupling into the top of your column.

Add a pipe elbow at the top of the reducing coupling. Don't spare the grease, for you want a good, tight fit to prevent your alcohol vapor from escaping.

A short length of pipe nipple should be screwed into the elbow.

You don't have to be a trained plumber or engineer to construct a junkyard still. If you can use a few basic hand tools and can follow directions, you're well on your way to success.

A heater hose from a junked car is attached to the other end of the pipe nipple and secured in place with a hose clamp. This completes the vapor line which leads from the still to the condenser.

There is nothing sacred about the size of pipe you use or the exact configuration and type of elbows. The author's model was hastily put together with what happened to be 3/4 inch pipe, street elbows, some close nipples and couplings. You might wish to come up with your own arrangement using 1/2 inch pipe. Keep in mind that

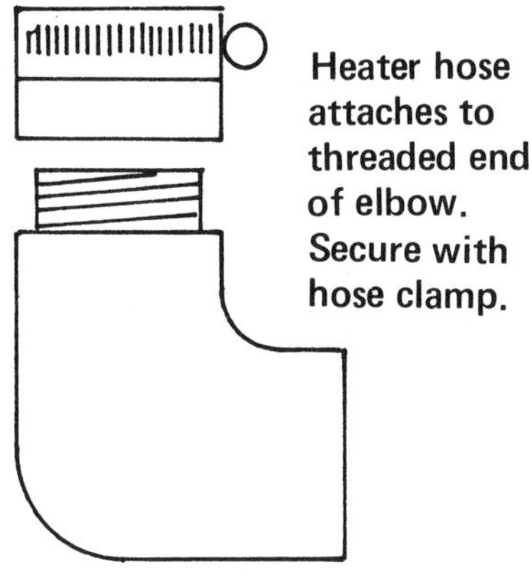

what is commonly known as 1/2 inch pipe has an over-all diameter (OD) of approximately 3/4 inch and that 3/4 inch pipe has an OD of approximately 1 inch. Since typical automotive heater hose has an inside diameter (ID) of 5/8 inch the 1/2 inch pipe might be a better choice. The same goes for the reducing coupling, elbow and nipple on top of the 2 inch pipe. Any good hardware store should have enough different pipe fittings to enable you to "tinker toy" your own version with ease.

You will note from the photographs that this condenser terminated with street elbows (the kind that have a female thread on one end and a male on the other). This allows the hose to be fitted over the male thread and clamped in place.

If you wish to use regular 90° elbows and close nipples, fine. A series of 12 inch nipples should now be attached with a double elbow arrangement on each end. Use enough zigzags to fill the tank (about six should do). The finished condenser will have two hoses attached — a long one to attach to the still and a shorter one to catch the alcohol with. This shorter hose passes through the hole in the bottom of the toilet tank where the float valve mechanism used to fit.

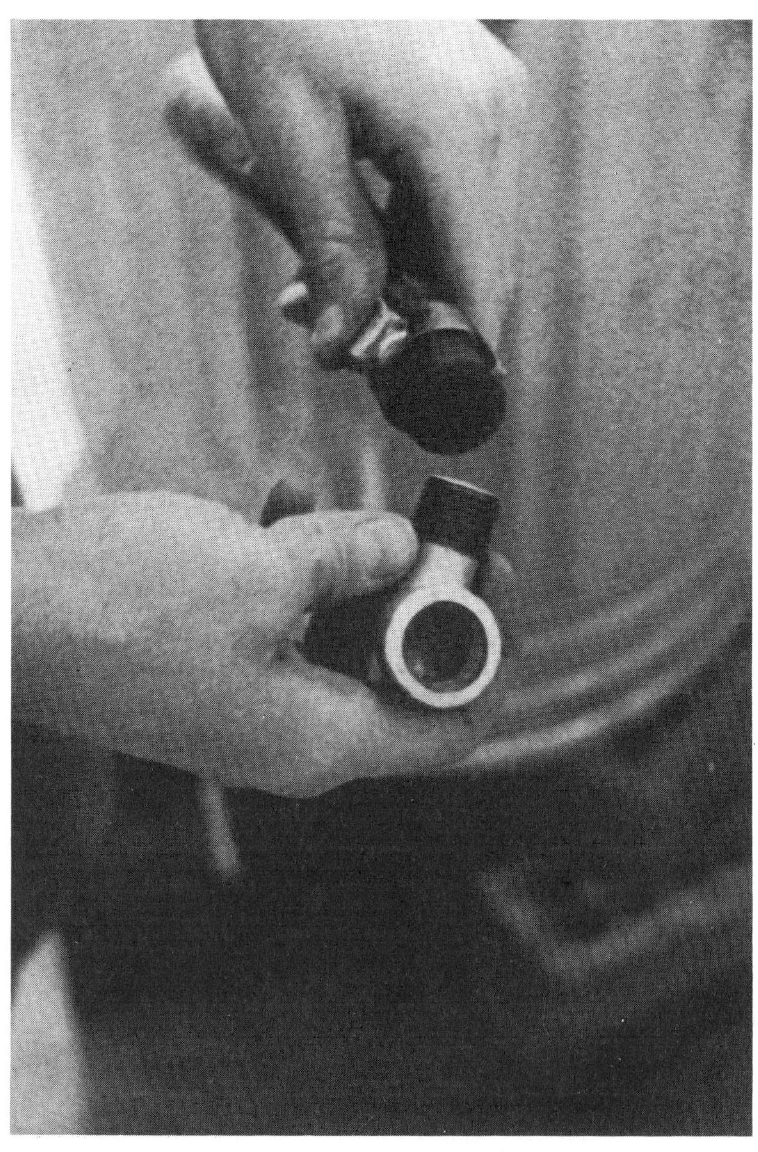

Insert the threaded end of a pipe elbow into the free end
of the vapor line (heater hose) and secure in place with
a hose clamp.

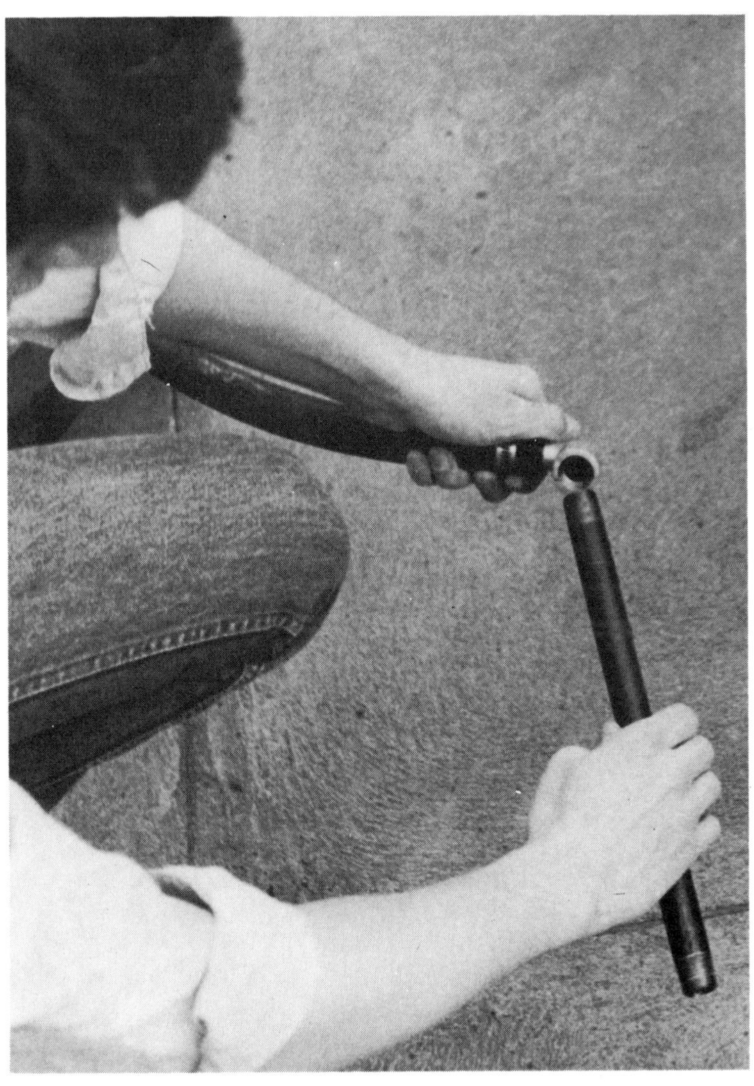

The cooling coil for your condenser may be constructed quickly and economically from 12 inch lengths of 3/4 inch pipe nipples and 3/4 inch elbows. This coil will easily fit inside a toilet tank, which holds the water that cools the alcohol vapor as it passes through the coil.

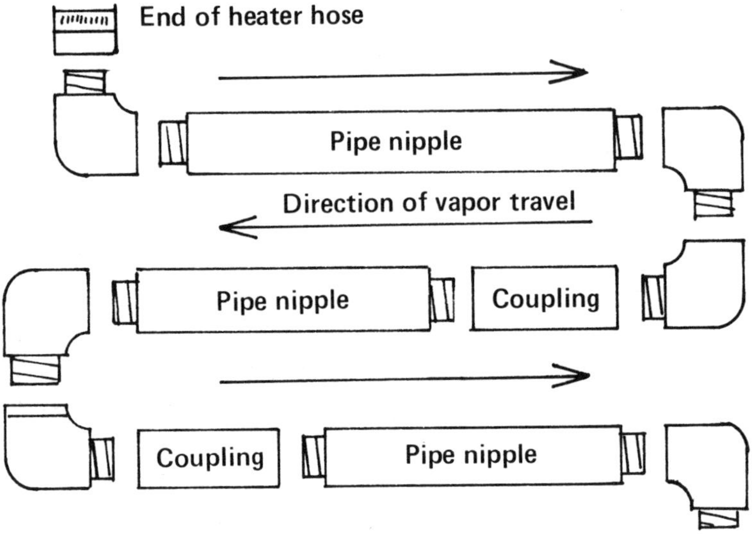

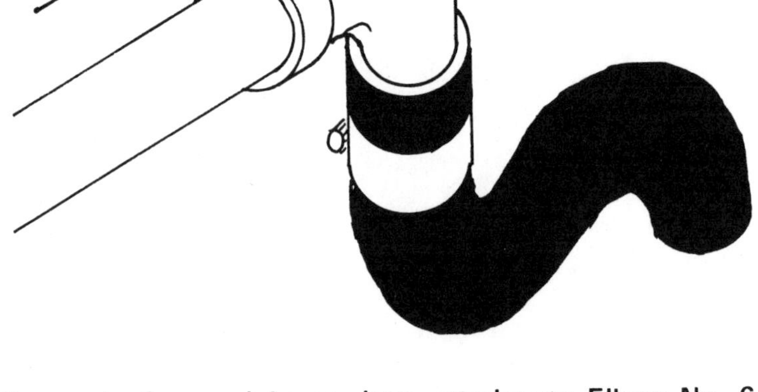

One end of second heater hose attaches to Elbow No. 6, other end passes through bottom of toilet tank and emerges at liquid alcohol collector.

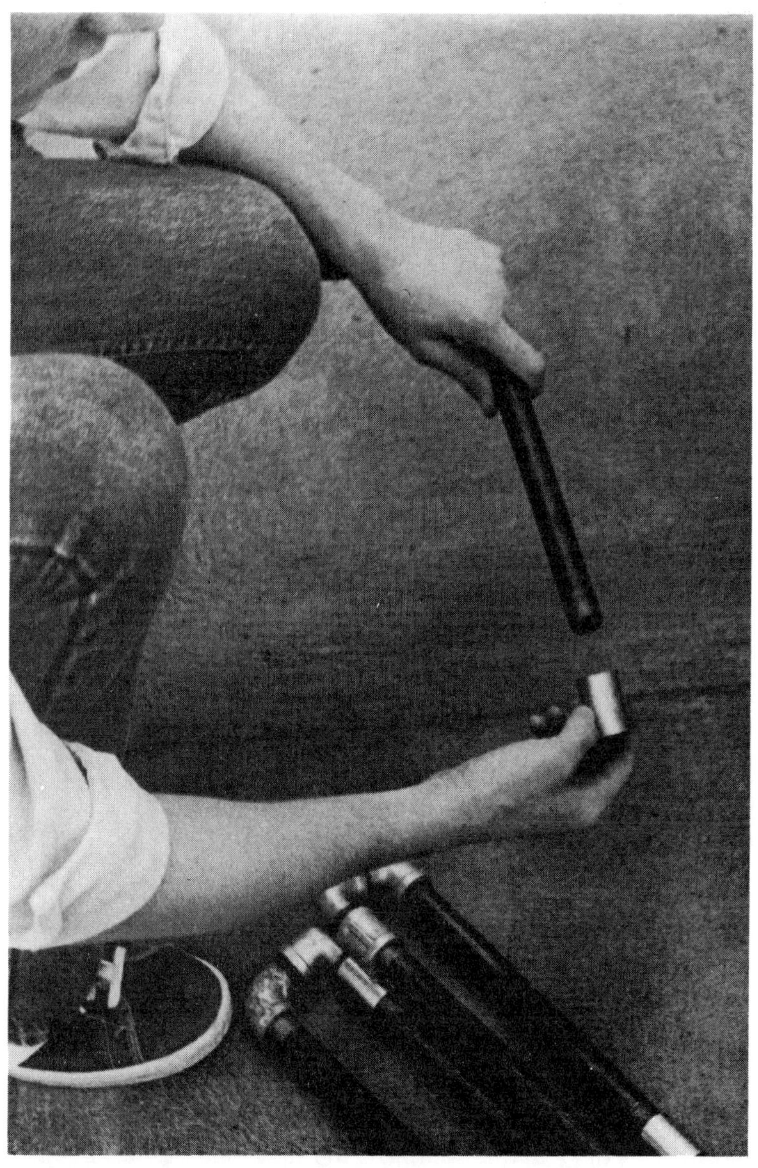

A short length of coupling is necessary to compensate for the two male ends of the pipe nipples.

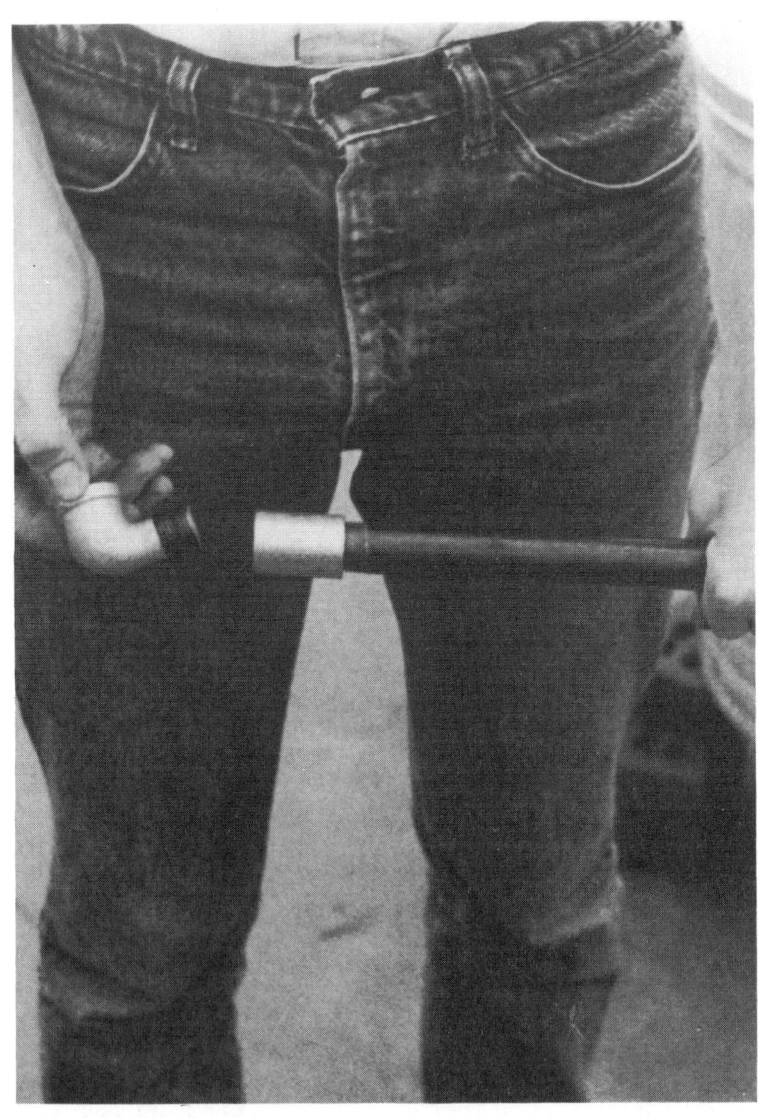

After attaching the coupling to the first pipe nipple, screw in a pipe elbow, then screw a second elbow into the first. Follow with a second length of pipe, then another coupling, then two more elbows, etc., etc., until your coil is completed.

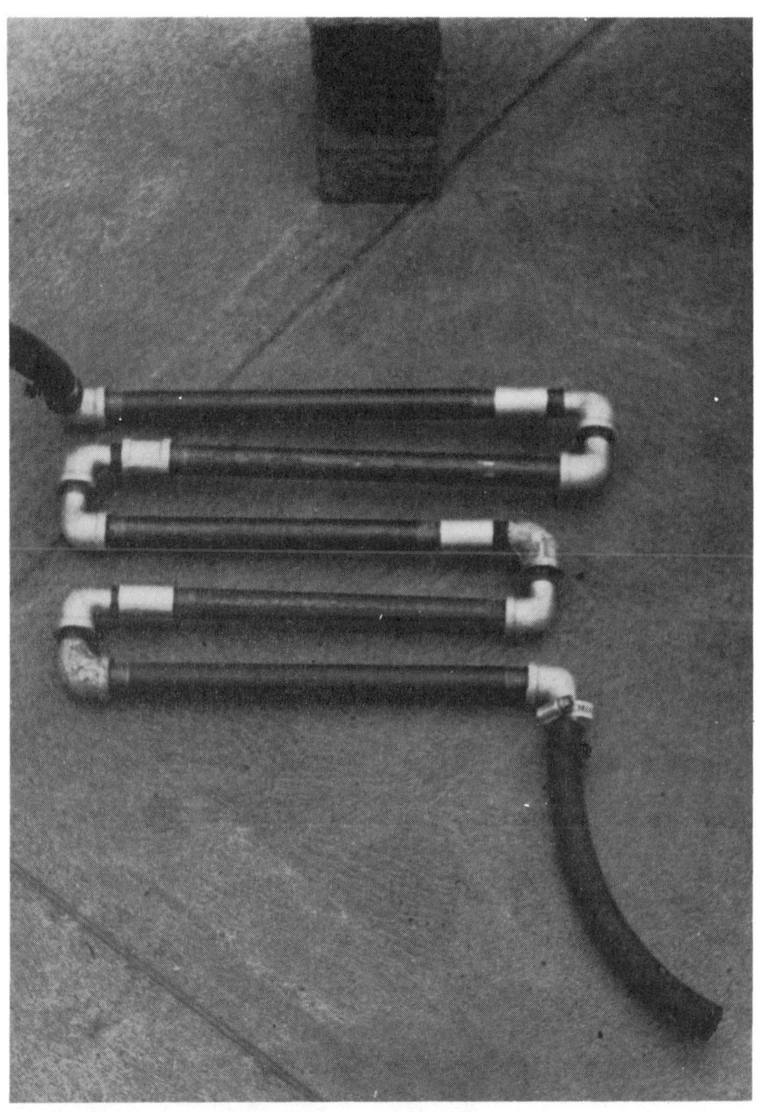

Your finished condenser coil should look like this. Alcohol vapor enters the coil through the hose at the top and slowly cools down as it passes through the coil, finally emerging in a liquid state from the hose at the bottom.

Plug the large hole in the bottom of your toilet tank condenser with a 2 inch pipe cap. Again, don't spare the grease
or gasket sealer.

The size of pipe and hose used on the model photographed made it self-sealing.

The large hole in the tank can be stopped up many ways. The way it was done below was by using a 2 inch close nipple, pipe cap, a large foam rubber washer (the one used to attach the tank to the toilet) and a 2 inch conduit jam nut. Use a generous amount of gasket sealer and carefully tighten this assembly in the hole. The remaining small holes (2 or 3) can be taken care of with screws and rubber washers.

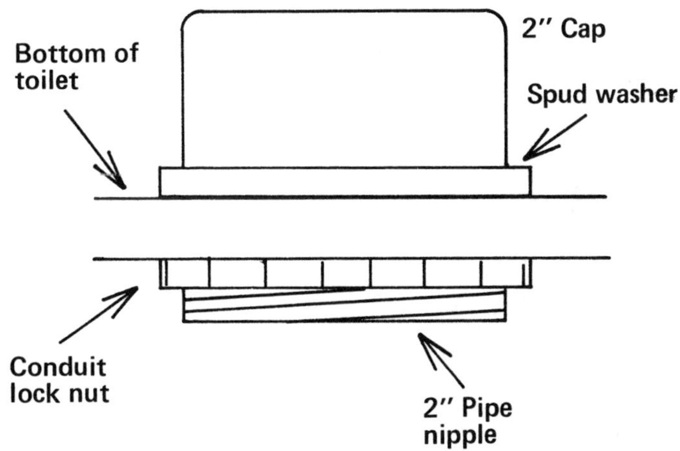

With the condenser installed, fill the tank with water and check for leaks. The same goes for the cooker. Run off a batch of distilled water before you try alcohol. You may have to remove the packing (rocks, marbles or whatever) from your column before water will distill.

The garden hose introduces cooling water into the tank while the warm water flows out the overflow. Try to keep the water at 60 degrees F.

If you want to get fancy try this: drill two holes — one at the bottom and one at the top of your column and insert two meat thermometers. Ideally, the thermometers should be held

Smaller holes around largest hole in the bottom of the toilet tank can be plugged up using machine bolts, rubber washers, lock washers and nuts of the proper size. The length of heater hose that is attached to the bottom of the coil will pass through the medium-sized hole at upper right.

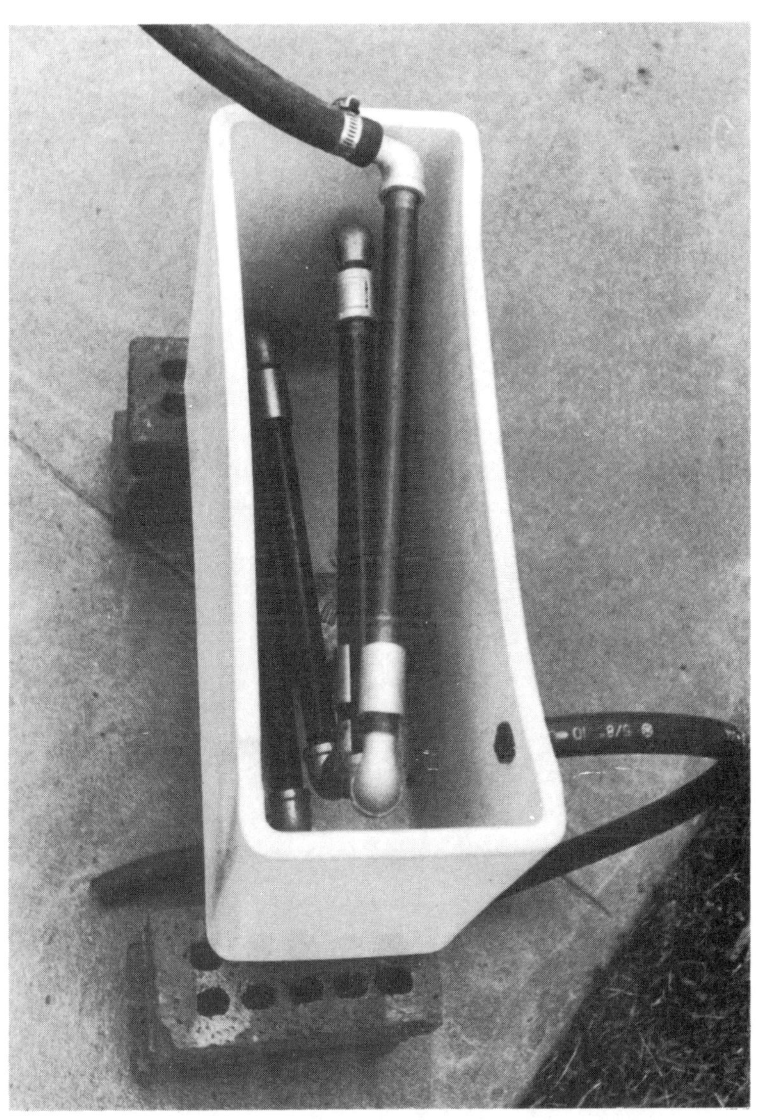

Pipe coil is cooled by water from an ordinary garden hose (not shown). The hole at the lower right formerly held the toilet's handle for flushing, now houses a short length of heater hose that serves as an overflow valve.

Your finished junkyard still should look something like this.

An alternate method of building a condenser is to use an old car radiator instead of a pipe coil and an old-fashioned wash tub instead of a toilet tank. Remember, you can substitute materials all along the way as long as you follow the basic principles covered within these pages.

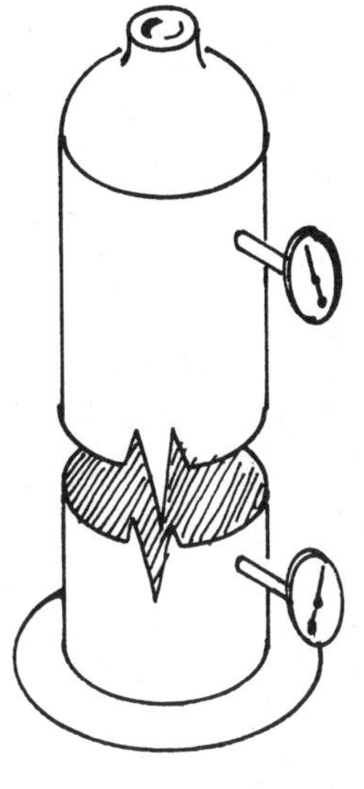

in place with rubber stoppers (ones with a hole in them) like are used in chemistry labs. When you begin cooking, try to keep the bottom thermometer close to 212 degrees F. and the top one at 173 degrees F.

Use a 55 gallon drum instead of a garbage can. Some have a bung hole that your column will screw right into. In this case the packing retaining screen can be held by using two reducing bushings small end to small end with a close nipple between them.

A WORD OF CAUTION - When using the garbage can cooker you have a natural safety feature. That is the way the lid fits on with the gasket material sealing it. Unless you mechanically fasten it or weight it down, the lid will simply raise up and dump excessive pressure. Not so with the 55 gallon drum unless pre-

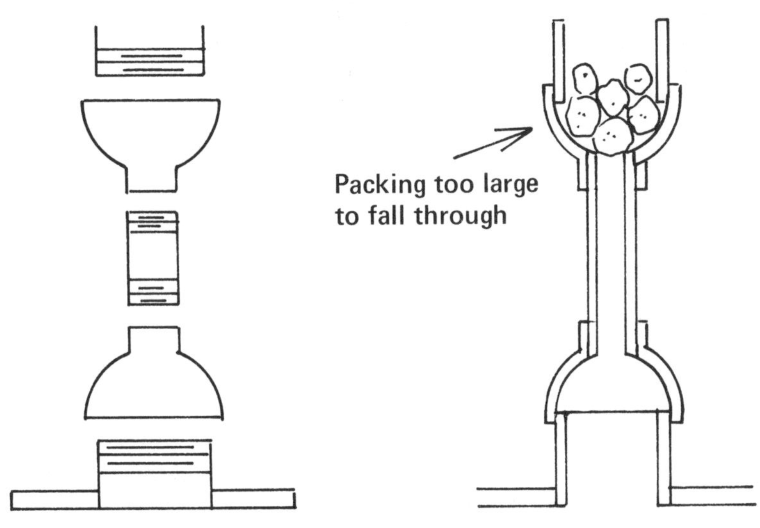

cautions are taken to insure that the column and condenser are never clogged up and that the fire under the drum is controlled to insure that the volume of steam does not exceed the flow capacity of the column—condenser arrangement.

Make sure you test the end product with a hydrometer when you drain it from the bottom of your cooker. It should register "0" after you have stripped all the alcohol from the mash.

Another area you may choose to be creative in is in the field of "condenser research." The air conditioning unit out of junked cars makes an excellent condenser. However, it will not fit in the back of a toilet. You will have to improvise another means of holding the cooling water.

A car radiator can be used but it's tricky. If you use a cross-flow radiator you will have to turn it up on one end.

Cross flow radiators are identifiable by the cooling tubes running from left to right.

A regular radiator works like this:

The hot steam (alcohol and water vapor) separates and travels down the tubes. There are so many tubes and they are so tiny that the cooling surface is equal to that of a coil with many turns.

Back in the 1930's one enterprising fellow tried to make drinking whiskey like this and either killed or blinded 41 people.

Finally, you may want to install an aspirator. What an aspirator does is reduce atmospheric pressure against the surface of a liquid. A liquid is a group of gas molecules held in place by the pressure of the blanket of air resting on the surface of the earth. This pressure changes with altitude but unless you live high in the mountains can be considered approximately 14.7 lbs. per square inch.

Aspirators come in many forms. Water powered. Electric vacuum pumps, etc. They all operate by creating a vacuum.

The boiling point of a liquid is reached when the heat (or energy) of the liquid (upward pressure) equals the atmospheric pressure (downward pressure).

If we can reduce the downward pressure, then it will require less upward pressure to boil a liquid. This is known as vacuum distillation.

There are other ways to reduce atomospheric pressure, including many ways probably only awaiting the discovery of some backyard junk collector. Just don't try for an excessive amount of vacuum or you will find out that 14.7 lbs. per square inch outside pressure will crush your cooker like an egg shell!

WARNING - DO NOT ATTEMPT TO DRINK ANYTHING DISTILLED IN THIS DEVICE. THE TIN SOLDER IN THE RADIATOR AND THE TIN IN THE GARBAGE CAN MAY UNDERGO CHEMICAL REACTIONS WITH THE ORGANIC COMPOUNDS IN YOUR ALCOHOL-PRODUCING MIXTURE AND FORM DEADLY POISONS!!!

Moonshine Kept Zeroes Flying

You wouldn't think a bunch of illiterate rice farmers from southeast Asia, working with barnyark junk, weeds, and some copper tubing, could produce better aviation fuel than Standard Oil — but they can.

And they did!

It is common knowledge that the Japanese main fighter plane, commonly referred to as the "Zero" could outfly a lot of our own fighter planes at the beginning of World War II. The history books tell us that the Zero was "lighter" than our planes and hence faster.

One major detail is always overlooked.

According to a recently released Department of Energy report, when American Marines began capturing Japanese airfields during the war, they found supplies of "aviation fuel" (so marked on the drums) that turned out to be grain alcohol, or ethanol.

The performance difference between gasoline and ethanol can be demonstrated in a passenger car just as readily as in airplanes.

Harold Jenks, president of Alcohol Fuel Injection of Ohio, Inc., Dayton, OH, demonstrated the performance difference to me in a 1970 six-cylinder Chevy Nova set up to run on either gasoline or alcohol. Top speed with both of us in the Nova on level ground, running on gasoline, was 90 mph. The two of us in the car had a collective weight of over

500 pounds (Harold is close to seven feet tall — I'm the pudgy one).

When we switched to alcohol, under the above conditions, the car accelerated past 105 and was still accelerating when we decided that red-line for this old Chevie had been reached. Our faith in alcohol as a motor fuel far exceeds our faith in Detroit tires and suspension systems!

The Zero had no special engine. If you get a chance to eyeball one at an antique aircraft fly-in you will notice the nine-cylinder radial engine in the Zero is almost a carbon copy of the Pratt & Whittney found in some of the older American planes.

It took our Armed Forces' bigger engines, thicker armor, electronic gun controls, fuel injection, supercharging and much more to overcome the Japanese "aviation fuel" — not to mention alcohol and water injection units that our fighter planes had to be equipped with.

Four Fodders For Ethanol Distillation

There are four basic groups of substances that can be processed into ethanol (as opposed to methanol).

Ethanol is commonly thought of as "farmer's alcohol" and is the basis for drinking alcohol.

Methanol, made from coal, is constantly confused with ethanol (probably deliberately) in testimony by oil company representatives before Congress. The fumes are toxic. It turns into formaldehyde in the human body (causing blindness or death), has only 73% of the BTUs ethanol contains, causes puddling in intake manifolds in gas engines since it vaporizes so poorly, and a host of other problems.

Ironically, 93% of all the industrial alcohol produced in the United States — and I'm referring only to industrial ethanol — is manufactured from petroleum. Ethanol from petroleum is distinguished from farmer's alcohol by the term, "fermentation alcohol." Chemically they are identical. That is, drinking liquor can be made from petroleum.

The other three basic groups are sugar crops, starch crops and cellulose. Obviously these are the crops used for the alcohol to make "gasohol." And these are the crops responsible for one gallon of fermentation alcohol removing the tax from 10 gallons of motor fuel — nine gasoline gallons and one alcohol gallon. In some states this means that 13 cents a gallon is saved by the oil refineries, a savings of $1.30 in taxes for every 10 gallons. At anything under $1.30, the alcohol is not only "free" to the oil companies, but they make a profit over and above normal. According to Harold Jenks, president of Alcohol Fuel Injection of Ohio, Inc., this is a "throw the farmers a fish" program to keep them locked into letting the oil companies control the production of fermentation alcohol.

Which, if permitted, has consequences just as serious and stifling as ever, especially the effect on the American farmer.

Contrived fuel shortages are operated through horizontal and vertical monopolies maintained by big oil. Of course, in all fairness to the oil companies, we must admit that the last oil shortage wasn't entirely their fault. David Rockefeller's Chase Manhatten Bank had loaned more money to the OPEC nations than they could pay back unless the price of crude was raised enough to pay the interest. That shouldn't surprise anyone. The United States Senate handed over the Panama Canal to protect the interests of Midland Marine Bank; said bank having loaned Omar Torrijos more money than he could ever afford to pay back. It happens all the time!!!

Then there is suppression of worthwhile fuel-saving inventions. If an engine became readily available tomorrow that used 20% of the fuel used by a comparable size-efficient standard car engine an immediate oil glut would be created. If there isn't one already.

One such engine is the Bourke. It has three moving parts. On the same load and speed, a typical Bourke uses one-half gallon of fuel in an hour, compared to two and one-half gallons for a Volkswagon engine.

There are governmental boondoggles at taxpayer expense, "Synthetic fuel" development, and all the other nonsense — like methanol from coal. The Department of Energy has already pointed out that processing methanol from coal will require five years of development and $500 million per plant investment, which is probably why oil company spokesmen tell our elected representatives so much about methanol. No farmer has that kind of money. Big oil can create another monopoly. On the other hand, a highly sophisticated fermentation alcohol plant producing two million gallons a year (and 10 million pounds of cattle feed) can be put together for slightly over $2 million, well within the range of many farm co-ops.

Let's review the basic crop groups.

Cellulose is the hardest to work with. Any time we want to ferment a crop of any sort we have to reduce it to a sugar, a very short carbon chain. Cellulose often consists of 3,000 sugar molecules stuck together. Before any of them can be converted to alcohol they have to be unstuck. The easiest way with cellulose is to simply grind the stuff up fine and dump it in a solution of sulfuric acid. At present, this process is too expensive. Considerable work is being done with enzymes taking the place of sulfuric acid at various colleges around the country, but nobody seems to have hit the jackpot yet.

Starch is the next hardest. It must be malted, a process too long to go into here, cooked at certain temperatures, and generally wet-nursed all along the way.

Sugars are fairly simple, since all they have to do is be put into solution. That is, if you go to the local supermarket and scavenge a bushel of rotten fruit. All you have to do is slice it up and dump it into a 55-gallon drum full of water between 63 and 90 degrees F. While at the supermarket pick up a package of baker's yeast and throw it into the drum along with the water and rotten fruit. If the water in the drum is kept at the correct temperature you should begin to see the fermentation process the next day. The yeast consumes the fruit sugar and excretes alcohol and carbon dioxide. The alcohol remains in solution and the carbon dioxide, being a gas, forms air bubbles, rises to the surface and escapes into the atmosphere. When you see the bubbles rise to the surface that is your indication that the fermentation process has actually started. Let it alone about three days, and then pour the clear liquid off into your cooker. The dregs or slop left behind can be converted into cattle feed.

What you wind up with in your cooker is only going to be 8 to 12% alcohol, hardly strong enough to get a buzz on, let alone run an engine.

Every time you re-distill it you should double the proof. With a crude moonshine still don't expect more than a one and one-half gallons per bushel yield of pure alcohol. Or a couple of gallons of 160 proof, quite adequate for motor fuel.

If you have trouble getting fruit to ferment you might check the pH factor of your water. Any pharmacist will sell you a small litmus paper kit with instructions. Fermentation does best in water that is slightly acid.

Sugar beets would probably be the most logical crop for alcohol production in the Midwest. Some folks attending a seminar asked how to work them up for fermentation. The answer was as follows:

— Shred the beets as finely as possible.

— Squeeze out the juice in some sort of a press (such as a cider prss).

— Shred a couple of green tops and dump them into the juice — this provides nitrogen for the yeast to grow on.

— Dump about 40 pounds of this mess into a 55-gallon drum.

— Pour warm water, slightly under 160 degrees F., up to three-fourths of the drum's capacity. If you go over 160 degrees , you will caramelize the beet sugar in the solution and wind up with a non-fermentable goo.

— Let the solution cool down to 70 degrees F. or there-abouts and add the yeast. Watch for the telltale carbon

dioxide bubbles the next day. If you add your yeast over
90 degrees F., it will die from the heat, under 63 it will
merely go dormant.

— After three days of fermentation pour off the clear
liquid into your cooker and distill it. Don't let your solution
ferment too long or you will wind up with vinegar.

A Final Thought

Whoever is responsible had better consider one item before the American farmer takes over the motor fuel business with farm-produced alcohol:

What are we going to do with all the byproduct?

For every bushel of corn it is possible to extract 2½ gallons of alcohol and 18 pounds of cattle feed known as distiller's dried grains (DDG). The balance of the fermentation product escapes into the atmosphere as carbon dioxide. Carbon dioxide can be turned into dry ice or the chemical used in some fire extinguishers, but it is generally a limited market.

The alcohol, of course, is no problem. Simply modify the carburetor on a normal automobile engine and pollution will be about a third of what it normally is with gasoline. Raise the compression ratio enough to take advantage of the full octane rating of alcohol (about 120), and pollution goes down to almost nothing. A car that has been getting 23 miles per gallon with gasoline will start at about 19 mpg with alcohol and eventually go to 22 mpg on alcohol as the stuff burns off the carbon in the heads. (Heat is energy, carbon is a heat absorber.) Take the same car and raise the compression ratio past what will work with gasoline (about 15 to 1; most current gasolines cause problems at anything after 9 to 1) and the miles per gallon should go over 30. Any time compression is raised one increment (like from 9 to 1 to 10 to 1), fuel consumption drops about 5%.

The problem is all that cattle feed. If it was pumped out of the bottom of the still all dried out and tied up in little plastic bags we could just store it and forget it. Unfortunately, life is never that simple.

Back in the 19th century the only way to dispose of hundreds of gallons of spent mash was to keep a feedlot for cattle handy. The leftover mash residue was simply dumped over ground-up corn cobs and poor quality hay. The stuff couldn't be dumped out on the ground, since it would contaminate the water supply. What happens is that some of it goes into suspension in the water, big chunks are easy to deal with, and the rest of it goes into solution where it forms not a mixture (like water and sand) but a chemical compound.

The primary disadvantage to this procedure (which is still followed in many cases, especially with large moonshine operations) is that production was always limited by the number of cows in the feedlot. A 500-cow feeding operation limited alcohol production to one-third of what it would have been for a 1,500-cow operation. Too bad if you get an order for 1,000 heifers and 15,000 gallons of alcohol the same day. You would certainly lose one sale or the other.

Later on the drum-type dryer was invented. It is still used in many places by many companies. It is not the best way to do things. A drum-type dryer requires a high internal temperature to get the job done and the result is often that the dried grains come out looking like charcoal (all burned up) and vitamin B is destroyed. Heat does that to B vitamins. Ideally the dried grains should be light colored.

Some of the coarser grains can be filtered out by passing the slop over a fine screen, tilted at a slight angle, and then dried in various ways — but there still is a lot of slop left. Incidentally, chickens and a few other barnyard critters will consume this stuff, too. Horses normally won't unless it has been completely dried out for them. Horses are such picky eaters!

The main thing you have to watch for is that you don't feed it straight. Distiller's dried grains have three times as much protein and vitamins as the crop it came from (three-fold concentration).

The most efficient way to produce DDG is with an evaporator room.

Mash that has been stripped of its alcohol content is pumped into a holding tank, kept heated if the stuff is going to be there more than three days or it will go sour, and from there into a centrifuge. Coarse particles are filtered out and sent in one direction while that in solution is sent off in another.

The solution is run through three "effect" stations which remove the bulk of the water and leave a solution that is 25% DDG and 75% water. The effect stations consist of an airtight compartment made by large metal plates joined together (the plates have holes in the middle for filtering), and a large vacuum chamber at the back end. Once the solution has reached 25% concentration level, it is pumped back in with the coarse particles, the whole mess is dried out, and it is put on a conveyor belt to a silo. From there it is sold and loaded into trucks or railroad cars.

An evaporator room does not take many people to run it — about two to a shift. If it weren't for the plates in the effect stations getting gummed up in about a week and having to be cleaned and rotated and all the leaks constantly springing in the extensive plumbing, the entire operation could be run by push button.

A normal size evaporator room can crank out 1,800 pounds of DDG an hour and run 24 hours a day. DDG currently is wholesaling for between a nickel and seven cents a pound. Of course, an evaporator room of this capability costs a million dollars to build. Ideally, farmers would form a cooperative to build one — and if they have their own distilleries, simply lay pipe from their distillery to the evaporator room and receive payment in DDG for their share.

And here is what we had better consider:

If (or should it be "when") we all start using alcohol instead of gasoline as a motor fuel, every time we burn one gallon of fuel we're going to have to dispose of more than seven pounds of DDG. To drive the point home a little further, let's say you take a trip and burn 10 gallons of gasoline. Had that been alcohol, you would have become responsible for disposing of more than 70 pounds of DDG. Multiply that by the millions of cars in this country and you can see we might have a problem.

Most people remember the political slogan used during the Great Depression of the late 20's "a chicken in every pot." We may be facing "two steers in every freezer" out of necessity. Then again, it might not be a problem.

The American farmer's takeover of the fuel business might be the beginning of the Millenium!

The material contained herein is devoted solely to the construction of a junkyard still and to some background material on ethanol alcohol, its past and potential future. For more detailed information on fermentation, more sophisticated stills and adapting internal combustion engines to run on alcohol, please refer to "Brown's Alcohol Motor Fuel Cookbook".

Notes

Notes

Other Books Available From Desert Publications

081 . . . . . Crossbows/ From 35 Years With the Weapon $11.95
206 . . . . . CIA Field Expedient Preparation of Black Powder........$11.95
221 . . . . . Evaluation of Improvised Shaped Charges $11.95
411 . . . . . Homeopathic First Aid $11.95
432 . . . . . Cold Weather Survival $11.95
435 . . . . . Homestead Carpentry $14.95
444 . . . . . Leadership Hanbook of Small Unit Ops $9.98
454 . . . . . Survival Childbirth $13.95
C-002 . . . How To Open A Swiss Bank Account $9.95
C-011 . . . Defending Your Retreat........$9.95
C-020 . . . Methods of Long Term Storage........$9.95
C-023 . . . Federal Firearms Laws $11.95
C-024 . . . Select Fire Uzi Modification Manual $13.95
C-025 . . . How to Build Silencers $6.95
C-027 . . . Firearms Silencers Volume 1 ....$11.95
C-029 . . . M1 Carbine Arsenal History........$7.95
C-032 . . . The Silencer Cookbook $9.95
C-038 . . . How To Build A Beer Can Morter $4.95
C-040 . . . Criminal Use of False ID $14.95
C-045 . . . Improvised Weapons of American Underground........$11.95
C-048 . . . Catalog of Military Suppliers $18.95
C-056 . . . Navy Seal Manual $29.95
C-065 . . . Poor Man's James Bond Vol 1 ....$34.95
C-113 . . . Submachine Gun Designers Handbook $19.95
C-114 . . . AR-15, M16 and M16A1 5.56mm Rifles $13.95
C-117 . . . Ruger Carbine Cookbook $11.95
C-120 . . . Shootout II........$14.95
C-122 . . . AK-47 Assault Rifle $15.95
C-123 . . . Combat Loads for Sniper Rifles $16.95
C-125 . . . Browning Hi-Power Pistols $13.95
C-126 . . . UZI Submachine Gun $13.95
C-127 . . . FullAuto Vol 1 Ar-15 Modification Manual $11.95
C-128 . . . Full Auto Vol 2 Uzi Mod Manual $11.95
C-130 . . . Full Auto Vol 4 Thompson SMG $11.95
C-131 . . . FullAuto Vol 5 M1 Carbine to M2........$11.95
C-135 . . . M-14 Rifle, The $12.95
C-136 . . . Fighting Garand Manual $13.95
C-137 . . . Ranger Training & Operations $19.95
C-138 . . . Emergency War Surgery $29.95
C-140 . . . Emergency Medical Care/Disaster........$14.95
C-143 . . . Survival Medicine $15.95
C-150 . . . Improvised Munitions Black Book Vol 1 ....$17.95
C-151 . . . Improvised Munitions Black Book Vol 2 ....$17.95
C-152 . . . Impro. Munitions Black Book Vol 3........$29.95
C-155 . . . Fighting Back on the Job $14.95
C-160 . . . Firearm Silencers Vol 2........$24.95
C-162 . . . Colt .45 Auto Pistol Manual $11.95
C-163 . . . Survival Evasion & Escape $19.95
C-165 . . . USMC Sniping $18.95
C-175 . . . Beat the Box $7.95
C-197 . . . How to Make Disposable Silencers Vol. 1 ....$18.95
C-198 . . . Expedient Hand Grenades $21.95
C-199 . . . U. S. Army Counterterrorism Training Manual $19.95
C-203 . . . Explosives and Propellants........$14.95
C-204 . . . Improvised Munitions/Ammonium Nitrate $13.95
C-205 . . . Two Component High Explosives Mixtures $14.95
C-207 . . . CIA Improvised Sabotage Devices $14.95
C-210 . . . Guerilla Warfare........$14.95
C-211 . . . FN-FAL Auto Rifles........$21.95
C-218 . . . Professional Homemade Cherry Bomb $15.95
C-219 . . . Improvised Shaped Charges $14.95
C-220 . . . Improvised Explosives, Use In Detonation Devices $14.95
C-227 . . . Springfield '03 Rifle Manual $15.95
C-229 . . . Lock Picking Simplified $9.95
C-230 . . . How to Fit Keys by Impressioning........$9.95
C-231 . . . Lockout -Techniques of Forced Entr $13.95
C-235 . . . How to Open Handcuffs Without Keys $11.95
C-244 . . . Camouflage $12.95
C-274 . . . M14 and M14A1Rifles and Rifle Marksmanship $23.95
C-284 . . . Thompson Submachine Guns $19.95
C-289 . . . M1 Carbine Owners Manual $13.95
C-290 . . . Clandestine Ops Man/Central America $11.95
C-324 . . . How to Make Disposable Silencers Vol. 2 ....$18.95
C-333 . . . USMC AR-15/M-16 A2 Manual $21.95
C-337 . . . Sten MKII SMG Construction Manual $24.95
C-356 . . . Map Reading and Land Navigation $19.95
C-365 . . . Survival Gunsmithing $14.95
C-372 . . . Poor Man's James Bond Vol 2 ....$34.95
C-386 . . . Self-Defense Requires No Apology $13.95
C-414 . . . Select Fire 10/22 ....$15.95
C-415 . . . Construction Secret Hiding Places $15.95
C-416 . . . Brown's Alcohol Motor Fuel Cookbook $24.95
C-417 . . . Canteen Cup Cookery $9.95
C-418 . . . Improvised Batteries/Detonation Devices $14.95
C-421 . . . Walther P-38 Pistol Manual........$13.95
C-422 . . . P-08 Parabellum Luger Auto Pistol........$14.95
C-423 . . . Sten Submachine Gun, The........$4.98
C-427 . . . How to Train a Guard Dog $16.95
C-429 . . . The Anarchist Handbook Vol. 1........$14.95
C-430 . . . Beretta - 9MM M9........$15.95
C-432 . . . Mercenary Operations Manual $5.95
C-439 . . . Improvised Lock Picks $13.95
C-455 . . . Modern Day Ninjutsu $17.95
C-509 . . . Survival Guns $24.95
C-512 . . . FullAuto Vol 8 M14A1 & Mini 14 ....$11.95
C-518 . . . L.A.W. Rocket System $14.95
C-528 . . . With British Snipers, To the Reich $29.95
C-539 . . . Dental Specialist........$49.95

C-558 . . . Food. Fur and Survival $17.95
C-561 . . . The Squeaky Wheel........$14.95
C-562 . . . Survival Chemist Book $17.95
C-564 . . . Medical Specialist $49.95
C-571 . . . U. S. Marine Corps Scout/Sniper Training Manual $27.95
C-610 . . . Aunt Bessie's Wood Stove Cookbook $9.95
C-622 . . . Defensive Shotgun........$14.95
C-625 . . . MAC-10 Cookbook $11.95
C-627 . . . Brown's Lawsuit Cookbook $26.95
C-628 . . . Can You Survive $16.95
C-633 . . . Hand to Hand Combat........$11.95
C-634 . . . USMC Hand to Hand Combat $9.95
C-635 . . . Hand to Hand Combat by D'Eliscue $7.95
C-636 . . . Prisons Bloody Iron $13.95
C-639 . . . Police Karate $17.95
C-646 . . . Science of Revolutionary Warfare $16.95
C-668 . . . Survival Shooting for Women $7.48
C-680 . . . Infantry Scouting, Patrol. & Sniping $16.95
C-694 . . . CIA Field Expedient Incendiary Manual $17.95
C-761 . . . How to Lose Your X-Wife Forever $24.95
C-763 . . . Vigilante Handbook $17.95
C-775 . . . Alcohol Distillers Handbook $24.95
C-776 . . . How to Build a Junkyard Still $15.95
C-777 . . . Engineer Explosives of WWI $7.48
C-794 . . . US Marine Corps Essential Subjects........$16.95
C-815 . . . US Marine Bayonet Training $9.95
C-818 . . . The Poisoner's Handbook $29.95
C-829 . . . The Anachist Handbook Vol. 2 ....$14.95
C-889 . . . Napoleon's Maxims of War $11.95
C-890 . . . Company Officers HB of Ger. Army $9.48
C-891 . . . German Infantry Weapons Vol 1 ....$18.95
C-894 . . . Shotguns, eight Gov. Types $17.95
C-896 . . . Op. Man. 7.62mm M24 Sniper Weapon $11.95
C-899 . . . US Army Bayonet Training $4.48
C-9002 . . German MG-34 Machinegun Manual $19.95
C-9012 . . Firearm Silencers Vol 3 ....$16.95
C-9013 . . HK Assault Rifle Systems $27.95
C-9047 . . SKS Type of Carbines, The........$17.95
C-9048 . . Rough Riders, The $24.95
C-9052 . . Private Weaponeer, The........$14.95
C-9057 . . Keys To Understanding Tubular Locks $11.95
C-9060 . . The Anarchist Handbook Vol. 3........$14.95
C-9081 . . Dirty Fighting $11.95
C-9082 . . Lasers & Night Vision Devices $29.95
C-9083 . . Water Survival Training........$8.95
C-9084 . . Assorted Nasties $29.95
C-9085 . . Ruger P-85 Family of Handguns $19.95
C-9101 . . Guide to Germ Warfare $14.95
C-9102 . . CIA Field Expedient Method for Explosives Preparation $14.95
C-9119 . . Surviving Global Slavery $16.95
C-9120 . . Military Knife Fighting $14.95
C-9127 . . H&R Reising Submachine Gun Manual $16.95
C-9131 . . Live to Spend It........$29.95
C-9134 . . Military Ground Rappelling Techniques........$13.95
C-9135 . . Smith & Wesson Autos $19.95
C-9137 . . Caching Techniques of U.S. Army Special Forces........$14.95
C-9138 . . USMC Battle Skills Training Manual $49.95
C-9139 . . Survival Armory $27.95
C-9156 . . Concealed Carry Made Easy $17.95
C-9164 . . The L'il M-1, The .30 Cal. M-1 Carbine $17.95
C-9170 . . Urban Combat $27.95
C-9178 . . Apocalypse Tomorrow $17.95
C-9198 . . MP40 Machinegun Manual $16.95
C-9199 . . Clear Your Record & Own a Gun........$24.95
C-9200 . . Sig's Handguns Manual $19.95
C-9210 . . Sniper Training $24.95
C-9212 . . Poor Man's Sniper Rifle $18.95
C-9221 . . The Official Makarov Pistol Manual $14.95
C-9224 . . The Butane Lighter Hand Grenade $12.95
C-9228 . . Unarmed Against the Knife $14.95
C-9229 . . Black Book of Booby Traps........$29.95
C-9239 . . Militia Battle Manual $22.95
C-9247 . . Black Book of Revenge $15.95
C-9253 . . Bankruptcy, Credit and You........$18.95
C-9255 . . The Sicilian Blade........$15.95
C-9259 . . How to Build Practical Firearm Silencers $16.95
C-9262 . . Glock's Handguns $19.95
C-9263 . . Heckler and Koch's Handguns........$19.95
C-9264 . . The Poor Man's Ray Gun $11.95
C-9265 . . The Poor Man's R. P. G. $22.95
C-9279 . . Credit Improvement & Protection Handbook $19.95
C-9281 . . Build Your Own AR-15 ....$19.95
C-9282 . . How to Live Safely in a Dangerous World........$19.95
C-9293 . . Bllack Book of Arson $24.95
C-9322 . . How to Build Flash/Stun Grenades $19.95
C-9342 . . Entrapment $19.95
C-9359 . . The FN-FAL Rifle Et Al. $21.95
C-9367 . . The Mental Edge, Revised $17.95
C-9373 . . Survival Bible by Duncan Long $59.95
C-9405 . . The Internet Predator $17.95
C-9406 . . The Criminal Defendants Bible $49.95
C-9430 . . Protect Your Assets $24.95
C-9432 . . How To Build Military Grade Suppressors........$21.95
C-9441 . . Mossberg Shotguns $24.95
C-9478 . . Poor Man's Primer Manual, The $24.95
C-9546 . . Agents HB of Black Bag Ops. $14.95
C-9556 . . Improvised Rocket Motors $7.48

Desert Publications
215 S. Washington Ave. Dept. - BK-301
El Dorado, AR 71730 USA
info@deltapress.com